YOUR KNOWLEDGE HAS VALUE

AF151034

Bibliographic information published by the German National Library:

The German National Library lists this publication in the National Bibliography; detailed bibliographic data are available on the Internet at http://dnb.dnb.de .

Imprint:

Copyright © 2011 GRIN Verlag, Open Publishing GmbH
Print and binding: Books on Demand GmbH, Norderstedt Germany
ISBN: 9783656211815

This book at GRIN:

http://www.grin.com/en/e-book/192147/synthesis-and-characterization-of-silver-nanoparticles-by-rhizopus-stolonifer

Afreen Banu, Vandana Rathod, E. Ranganath

Synthesis and characterization of silver nanoparticles by "Rhizopus stolonifer" and its activity against multi-drug resistant "Escherichia coli" and "Staphylococcus aureus"

GRIN Publishing

GRIN - Your knowledge has value

Since its foundation in 1998, GRIN has specialized in publishing academic texts by students, college teachers and other academics as e-book and printed book. The website www.grin.com is an ideal platform for presenting term papers, final papers, scientific essays, dissertations and specialist books.

Visit us on the internet:

http://www.grin.com/

http://www.facebook.com/grincom

http://www.twitter.com/grin_com

Synthesis and characterization of silver nanoparticles by *Rhizopus stolonifer* and its activity against multidrug resistant *Escherichia coli* and *Staphylococcus aureus*

Vandana Rathod, Afreen Banu, E.Ranganath

Department of Microbiology, Gulbarga University, Gulbarga-585106, Karnataka, India

No conflict of interest was reported by the authors of this article.

*Corresponding author: Afreen Banu, Research Scholar, Dept of Microbiology, Gulbarga University, Gulbarga-585106, Karnataka, India.

Complete manuscript word count including body text and figure legends: **2,618**
Number of figures: **11**
Number of references: **19**

Abstract

This study reports the extracellular synthesis of silver nanoparticles by *Rhizopus stolonifer* and its efficacy against multidrug resistant (MDR) *E.coli* and *S.aureus* isolated from Khwaja Bande Nawas Hospital, Gulbarga, Karnataka. Synthesis of silver nanoparticles (AgNPs) was carried out by using fungal filtrate of *R.stolonifer* and an aqueous solution of AgNO₃. The characterization of AgNPs was made by UV-Visible absorption Spectroscope, Scanning Electron Microscope and Energy Dispersed Spectroscope (SEM-EDS), Transmission Electron Microscope (TEM), Fourier Transform Infrared (FTIR) spectroscopy and Atomic Force Microscope (AFM). TEM micrograph revealed the formation of spherical nanoparticles with size ranging between 3 to 20 nm. Atomic force microscopy gives the three dimensional structure of the particles. The presence of proteins was detected by FTIR spectroscopy. Three dimensional structure of AgNPs was studied by AFM. AgNPs produced by *R.stolonifer* gave good antibacterial activity against clinical isolates which were multidrug resistant. Here we report the efficacy of mycogenic metal nanosilver against MDR strains which is difficult through conventional chemotherapy.

Keywords: *R. stolonifer*, silver nanoparticles, MDR-strains.

1. Introduction

Nanotechnology is the ability to observe, measure, manipulate and manufacture things at the nanometre scale. A nanometre (nm) is a unit of length 10^{-9}, or a distance of one-billionth of a meter. That's very small. At this scale you are talking about the size of the atoms and molecules. Recently, scientist has endeavoured the use of microorganisms as possible eco-friendly nano factories for the synthesis of metallic nanoparticles [1]. An important aspect of nanotechnology is the development of toxicity-free synthesis of metal nanoparticles which is a great challenge. Where as chemical synthesis of nanoparticles have adverse effect due to the absorbance of toxic chemicals on the surface. Green synthesis provides advancements over chemical and physical methods as it is environment friendly, cost effective, easily scaled up for large scale synthesis and biological method does not require high pressure, energy and toxic chemicals.

New multidrug resistant strains of bacteria have become a serious problem in public health. The emerging resistances in bacteria and high cost of advanced antimicrobial drugs have encouraged researchers to search for effective and economically viable broadly applicable drugs [2]. It has been known for long time that silver ions are highly toxic to a wide range of bacteria and silver based compounds have been used extensively in bactericidal applications. Silver has one advantage of having broad antimicrobial activities against gram negative and gram positive bacteria. AgNPs are the most effective preparation of silver because of the high surface/volume fraction resulting in a large proportion of silver atoms in direct contact with their environment [2]. It can be expected that the high specific surface area and high fraction of surface atoms of AgNPs will lead to high antimicrobial activity compared to bulk Ag metal [3]. The purpose of this study was to examine the antibacterial activity of silver nanoparticles against multidrug resistant *S.aureus* and *E.coli.*

The objective of the work is to synthesize silver nanoparticles from *R. stolonifer* and to study the antibacterial activity of the biosynthesized silver nanoparticles on gram-positive and gram-negative strains isolated from Khwaja Bande Nawas hospital, Gulbarga. The MDR-strains of *E.coli, and S.aureu* were selected for antibacterial study with the SNPs produced by *R.stolonifer.*

2. Materials and Methods

2.1. Biosynthesis of silver nanoparticles

Fungal isolates from soil were inoculated in Malt Glucose Yeast Peptone (MGYP) broth [4] containing yeast extract and malt extract-0.3% each, glucose-1%, peptone-0.5%, at 40°C, in shaking condition (180 rpm) [4]. After incubation of 72h the biomass was filtered and then extensively washed with distilled water to remove the medium components. This biomass was taken into flasks containing 100 ml distilled water and were incubated at the aforesaid condition. After 72h the biomass was filtered again, (Whatman filter paper No.1) the fungal filtrate was used further. Aqueous solution of AgNO3 (1mM AgNO3 of final concentration) was mixed with fungal filtrate and the flasks were agitated at 40°C. periodically, aliquots of only those isolates which showed colour change from yellow to brown were subjected for UV-Visible absorption spectrophotometric study. Control (without silver ions) was also run along with the experimental flasks.

2.2. Characterization of silver nanoparticles

Synthesis of AgNPs was characterized by UV-Visible absorption Spectrophotometer (T90/T90+ double-beam) with a resolution of 1 nm, which is one of the important technique to verify the formation of metal nanoparticles provided surface plasmon resonance exists for the metal [5]. To detect silver nanoparticle the absorption range is 400 to 450 nm [6]. This surface Plasmon resonance is caused by the coherent oscillation of the free conduction electrons induced by light. The confirmation for the synthesis of elemental silver was made by energy dispersed spectroscope. Transmission electron micrograph pattern were recorded on a carbon-coated copper grid on a Hitachi-H-7500 machine, sample preserved for over 6 months, synthesized by treating silver nitrate solution with cell free filtrate of *R.stolonifer*. The interaction between protein and AgNPs was analysed by Fourier transform-infrared spectroscope (JASCO FT/IR-3500). Three dimensional structures of biosynthesized silver nanoparticles were observed by atomic force microscopy.

2.3. Source of microorganisms

Two MDR-strains of *E.coli* and *S.aureus* from Khwaja Bande Nawas Hospital were used to study the antibacterial efficacy of silver nanoparticles.

2.4. Analysis of the Antibacterial activity of silver nanoparticles

The effect of silver nanoparticles on gram-positive and gram-negative bacteria was investigated by culturing the organisms on Luria-Bertani (LB) agar (10^6 colony forming units (CFU) of each strain per plate) supplemented with nanoparticles of 0.5, 1, 1.5, 2, 2.5µg/ml. Plates without silver nanoparticles were used as controls. Plates were incubated for 24h at 37°C the number of colonies was counted. The counts on three plates corresponding to a particular sample were averaged.

To examine the MIC of AgNPs and the growth curve of *E.coli* and *S.aureus* to AgNPs, different concentration of nanosilver 0.5, 1, 1.5, 2, 2.5 µg/ml was added in LB medium. Each bacterium culture (*S.aureus* and *E.coli*) was controlled at 10^5-10^6 CFU/ml and incubated at 37°C. To establish the antibacterial activity of nanosilver on bacterial growth, the MIC of AgNPs was determined by optical density of the bacterial culture solution containing different concentrations of nanoparticles after 24h.

3. Result

R. stolonifer (Fig.1) was used for the synthesis of silver nanoparticles from aqueous solution of AgNO3. The colour change of the fungal filtrate was noted by visual observation. The appearance of brown colour solution clearly indicates the formation of silver nanoparticles [7, 8]. A series of typical UV-Visible absorption spectra of the reaction solution were recorded at every 24h can be observed in Fig.2. All the spectra exhibit an intense peak at 422 nm corresponding to the surface plasmon resonance frequency of silver nanoparticles. This event clearly indicates that the reduction of the ions occur extracellularly through reducing agents released in to the solution by fungi. Scanning electron microscope clearly shows the biosynthesis of well dispersed nanosilver particles by *R.stolonifer*. EDS analysis gives the additional evidence for the reduction of silver nanoparticles to elemental silver. The optical absorption peak is seen approximately at 3kev. A representative TEM micrograph shows the AgNPs size ranges between 5 nm to 30 nm. FT-IR spectrum shows the bands at 1633(3) and 1554(4), 1423(5) cm^{-1}. AFM gives clear three dimensional picture of the nanoparticles, the height and width of the particle is measured (5 nm) using the software. Silver nanoparticles at the concentration of 20µg/mL and 25µg/mL were effective against *E.coli* and *S.aureus* respectively.

4

4. Discussion

4.1. UV-Visible absorption Spectroscopy

UV-Visible absorption spectroscopy is one of the most widely used technique for structural characterization of silver nanoparticles. The colour change was caused by the surface plasmon resonance of silver nanoparticles in the visible region [9]. Silver nanoparticles are known to exhibit size and shape dependent surface plasmon resonance bands which are characterized by UV-Visible absorption spectroscopy [10]. Silver nanoparticles showed maximum absorbance at 422 nm, implying that the bio reduction of the silver nitrate has taken place following incubation of the AgNO3 solution in the presence of cell-free extract. Our results are correlating with the reports of Sadowski, and Maliszwaska (2009) with the fungus *Penicillium*. It is reported that the absorption spectrum of spherical silver nanoparticles presents a maximum between 420 nm and 450 nm [6].

4.2. SEM-EDS

Scanning electron microscope clearly shows the biosynthesis of well dispersed nanosilver particles (Fig. 3) by *R.stolonifer*. EDS analysis gives the additional evidence for the reduction of silver nanoparticles to elemental silver. The optical absorption peak is seen approximately at 3kev, which is typical for the absorption of metallic silver nanocrystals due to surface plasmon resonance, which confirms the presence of nanocrystalline elemental silver. Spectrum shows strong silver signal along with weak oxygen and carbon peak, which may be originate from the biomolecules that are bound to the surface of nanosilver particles can be seen in Fig.4.

4.3. Transmission electron microscopy

A representative TEM micrograph recorded from drop coated film of a silver nanoparticles sample preserved for over 6 months, this has been deliberately done to study the effect of ageing on the size of the particles. The AgNPs are spherical in shape. All the particles are well separated and no agglomeration was noticed can be observed in Fig.5. Biosynthesized AgNPs size ranges between 5 nm to30 nm. The process of growing silver nanoparticles comprises of two key steps: (a) bioreduction of $AgNO_3$ to produced silver

5

nanoparticles and (b) stabilization and/or encapsulation of the same by suitable capping agents [11]. It is suggest that the biological molecules could possibly perform the function for the stabilization of the AgNPs. Silver nanoparticles synthesized by this route are fairly stable even after prolonged storage. This may be concluded that there is not much agglomeration of the AgNPs even after preserving the colloidal solution for extended periods. Fig.6 clearly represents the stability of AgNPs even after 6 months of storage.

4.4. Fourier Transform Infrared spectroscopy

The aim of IR spectroscopic analysis is to determine chemical functional groups in the sample. The amide linkages between amino acid residues in polypeptides and proteins give rise to well known signatures in the infrared region of the electromagnetic spectrum. Different functional groups absorb characteristic frequencies of IR radiation. Thus, IR spectroscope is an important and popular tool for structural elucidation and compound identification. Fig.7 shows the bands at 1633(3) and 1554(4) cm^{-1} are identified as the amide I and amide II and arises due to carbonyl stretch and –N-H stretch vibrations in the amide linkages of the proteins, respectively. The band at 1423(5) cm^{-1} is assigned to methylene scissoring vibration from the protein in the solution. Overall the observation confirms the presence of protein in the sample which coat covering the silver nanoparticles known as capping proteins. Capping protein stabilizes the metallic nanoparticle and prevents agglomeration in the medium. This study gives the evidence of formation and stabilization of silver nanoparticles in the aqueous medium by using biological molecules.

IR spectroscopic study has confirmed that the carbonyl group from amino acid residues and peptides of proteins has the stronger ability to bind metal, so that the proteins could most possibly form a coat covering the metal nanoparticles to prevent agglomeration of the particles and stabilizing in the medium. This evidence suggests that the biological molecules could possibly perform the function for the formation and stabilization of the AgNPs in the aqueous medium.

4.5. Atomic force microscope

For more information about the biosynthsized silver nanoparticles the sample was subjected to atomic force microscopic study. Fig. 8a shows the particles which are spherical

6

in shape, smooth surface and monodispersed in nature under optimized condition for the production of silver nanoparticles. The topography of the picture shows the particles from three different places seen in Figure 8b. The height and width of the particle is measured (5 nm) using the software.

4.6. The antibacterial activity of silver nanoparticles

Bbiosynthesized nanosilver showed excellent antibacterial activity against multidrug resistant test strains. In case of MDR *E.coli* 40% inhibition in bacterial growth was observed in plates supplemented with 0.5µg/ml of nanoparticles. The inhibition increased to 80% in plates with 1µg/ml of nanosilver particles, where as the concentration of 1.5µg/ml ensured complete inhibition of bacterial growth can be seen in Fig.9. In case of *S.aureus* 70 to 75% inhibition in growth was observed in plates supplemented with 1µg/ml of nanoparticles, where as 1.5µg/ml or high concentration of nanoparticles elicited 100% inhibition in growth of bacteria shown in Fig.9.

In further experiments, gram negative and gram positive bacteria were inoculated in LB liquid media supplemented with AgNPs. Fig.10 shows the increasing concentration of nanoparticles progressively inhibited the growth of MDR test strains. The concentration of 1.5µg/ml was found to be strongly inhibitory for *E.coli. S.aureus* showed complete inhibition of bacterial growth at 2µg/ml depicted in Fig.11. Biosynthesized AgNPs were found to have a good antibacterial activity on the growth of gram negative as well as on gram positive bacteria.

Antibacterial activity of silver ions is well known; however, the bactericidal mechanism is only partially understood. It has been studied that ionic silver strongly interacts with thiol groups of vital enzymes and inactivate them [12]. Experimental evidence shows that DNA loses its replication ability once the bacteria have been treated with silver ions [12]. Most importantly silver attacks a broad range of targets in the microbes, so it is difficult for them to develop resistance against silver, because this would require developing a host of mutations to protect themselves [13]. Nanosilver is known to be nontoxic to human cells to the tune of 350µg/day [13].

Due to its strong toxicity to a wide range of microorganisms, silver based compounds have been used extensively in many bactericidal applications, which includes, wound dressings [14], dental materials [15], stainless steel materials [16], human skin [17], because of such wide range of applications, a simple eco-friendly, cost-effective technique have been proposed for the synthesis of stabilized silver nanoparticles. The biosynthesized silver nanoparticles with a narrow particle size are more effective antibacterial agents because of high surface volume so that a large proportion of silver atoms are in direct contact with their environment [19].

Here we report the efficacy of mycogenic metal nanosilver to kill MDR strains which is difficult through the conventional chemotherapy. It is reasonable to state that the binding of particles to the bacteria depends on the surface area available for interaction. Smaller particles having the larger surface area available for interaction will give more bactericidal effect than the larger particles. The gram negative bacteria have a layer of lipopolysaccharide at the exterior, followed by a thin layer of peptidoglycan negative charges on the lipopolysaccharides are attracted towards weak positive charges available on silver nanoparticles and disturb its power function such as permeability and respiration [17]. Nanosilver may also penetrate inside the bacteria and cause damage by interacting with phosphorus and sulfur-containing compounds such as DNA [18]. The results of this study clearly demonstrated that the colloidal silver nanoparticles inhibited the growth and multiplication of the test organisms. Such high antibacterial activity was observed at very low concentration of nanosilver 20μL (1μg).

However, synthesis of nanoparticles using *R.stolonifer* can potentially eliminate the problem of chemical agents, which may have adverse effects in its application, thus making nanoparticles more biocompatible. This process of the production of silver nanoparticles is environmental friendly as it is cost-effective and free from any solvents and toxic chemicals. The filamentous fungi are easy in handling and also easily amenable on a large scale production. The nanoparticles synthesized using fungi present good monodispersity and stability. The potential application of nanoparticles in different fields has revolutionized the health care textile and agricultural industry. The filamentous fungi are easy in handling and also easily amenable on a large scale production. In conclusion, we have reported a simple biological way for synthesizing the silver nanoparticles.

References

[1]. Fayaz.A.M, Biogenic synthesis of silver nanoparticles and their synergistic effect with antibiotics: a study against gram-positive and gram-negative bacteria. Journal of Nanomedicine: Nanotechnology, Biology and Medicine 2010; 6, 103-109.

[2]. Ingle.A, Gade.A, Mycosynthesis of Silver nanoparticles using the fungus *Fusarium acuminatum* and its activity against some Human Pathogenic Bacteria. Current Nanoscience 2008; 4, 141-144.

[3]. Cho.K.H, Park.J.E, The study of antimicrobial activity and preservative effect of nanosilver ingredient. Electrochimica Acta 2005; 51, 956-960.

[4]. Karbasian M., Atyabi SM., Siyadat SD., Momen SB and Norouzian D, Optimizing nano-silver Formation by *Fusarium oxysporium* PTCC 5115 Employing Response Methodology. Am.J.Agri. and Biol 2008; 3(1), 433-437.

[5] Basavaraja. S, Balaji. S.D, Arunkumar.L, Rajasab.A,H Venkataraman.A, Extracellular biosynthesis of silver nanoparticles using the fungus *Fusarium semitectum*. *Materials Research Bulletin* 2008; 43, 1164-1170

[6]. I Maliszewska., Z Sadowski, Synthesis and antibacterial activity of silver nanopartilces. Journal of Physics: Conference series 2009; 146, 012024.

[7]. Mukherjee P ,Ahmed A, Mandal D, Senapati S, sainkar SR ,Khan MI, et al, Bioreduction of $AuCl_4^-$ Ions by the Fungus, *Verticillium sp.* and Surface Trapping of the Gold Nanoparticles Formed. Angew Chem Int Ed 2001a; 40, 3585-3588.

[8]. Mukherjee P., Ahmad A., Mandal D., Senapati S., Sainkar SR., Khan MI R., et al, Fungus-Mediated Synthesis of Silver Nanoparticles and Their Immobilization in the Mycelial Matrix: A Novel Biological Approach to Nanoparticle Synthesis. Nano Lett. 2001b; 1, 515-519.

[9]. Varshney.R, Mishra.A.N, Bhadauria.S, Gaur.M.S, A novel Microbial route to synthesize silver nanoparticles using fungus Hormoconis resinae. Digest journal of Nanomaterials and Biostructures 2009; 4(2), 349-355.

[10]. Xie.J, Lee.J.Y, Daniel I. C, and Ting.Y.P, Silver Nanoplates: From Biological to Biomimetic Synthesis. Am Chem Soc 2007; 1(5), 429–439.

[11]. Mukherjee.P, Roy.M, Mandal.B.P, Green synthesis of highly stabilized nanocrystaline silver particles by a non-pathogenic and agriculturally important fungus T. asperellum. Nanotechnology 2008; 19, 075103(7pp).

[12]. Morones.J.R, Elechiguerra.J.L, The bactericidal effect of silver nanoparticles. J. Nanotechnology 2005;16, 2346-2353.

[13]. Patil.S.S, Dhumal.R.S, Synthesis and Antibactrial Studies of Chloramphenicol Loaded nano-silver against *Salmonella typhi*. Synthesis and Reactivity in Inorganic, Metal-organic and Nanometal Chemistry 2009; 39, 65-72.

[14]. Panacek.A, kvitek.L, Prucek.R, Silver Colloid Nanoparticles: synthesis, and their Antibacterial Activity. J. Phys. Chem.B 2006; 110, 16248-16253.

[15]. Ohashi.S., Saku.S., Yamamoto K, Antibacterial activity of silver inorganic agent YDA filler. J.Oral Rehabil 2004; 31,364-367.

[16]. Bosetti. M., Masse A., Tobin E., Cannas M., Silver coated materials for external fixation devices: in vitro biocompatibility and genotoxicity. Biomaterials 2002;23,887-892(6).

[17]. Gauger A., Mempel M., Schekatz A., Schafer T., Ring J., Abeck D, Silver-Coated Textiles Reduce *Staphylococcus aureus* Colonization in Patients with Atopic Eczema. Dermatology 2003; 207, 15-21.

[18] Rai.M, Yadav.A and Gade.A, Silver nanoparticles as a new generation of antimicrobials j.biotechadv 2009; 27(1), 76-83.

[19]. Chen. X, and Schluesener. H, Nanosilver, a nanoproduct in medical application. Toxicol. Lett 2008;176, 1-12.

Figure Legends

Fig.1. *R.stolonifer* on PDA

Fig.2. UV-Visible absorption spectroscopy of silver nanoparticles of *R.stolonifer* at different time interval

Fig.3. SEM micrograph showing well dispersed silver nanoparticles

Fig.4. EDS of AgNPs produced by *R.stolonifer*

Fig.5. TEM image show Silver nanoparticles Synthesized by *R.stolonifer*

Fig.6. Absorption spectra of silver nanoparticles recorded one week after the synthesis and after six months.

Fig.7. FT-IR spectra recorded from a drop-coated film of silver nanoparticles Synthesized by *R.stolonifer*

Fig.8a. AFM picture of the sample

Fig.8b. AFM shows the three dimensional image of the silver nanoparticles.

Fig.9. Bacteria grown on agar plates at different concentrations of silver nanoparticles (a) *E.coli* (b) *S.aureus*. In each figure the concentration of silver nanoparticles are as follows: upper left, 0 µg/ml, upper middle 0.5µg/ml, upper right 1µg/ml, bottom left 1.5µg/ml, bottom middle 2µg/ml, and bottom right 2.5µg/ml. upper left plate, agar without nanoparticles and without bacterial inoculation.

Fig.10. MIC of biosynthesized AgNPs against test strains

Fig.11. Bacterial growth curve in LB media at different concentration of AgNPs

(a) *E.coli* (b) *S.aureus*

Figures

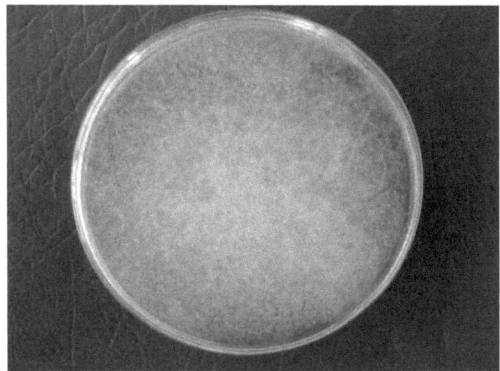

Fig.1. *R.stolonifer* on PDA

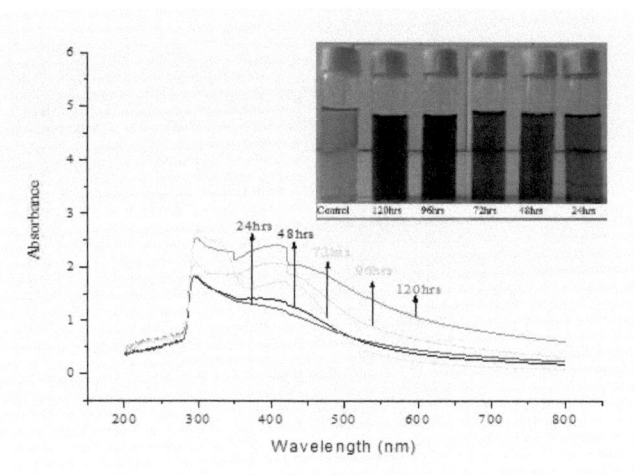

Fig.2. UV-Visible absorption spectroscopy of silver nanoparticles of
   R.stolonifer at different time interval

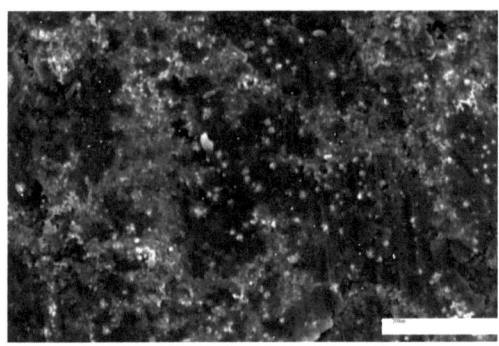

Fig.3. SEM micrograph showing well dispersed silver nanoparticles

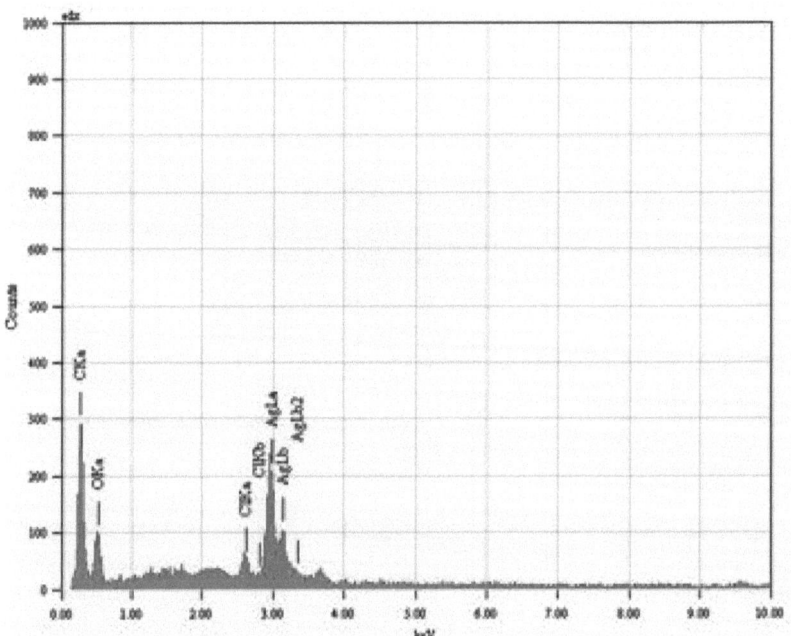

Fig.4. EDS of SNPs produced by *R.stolonifer*

13

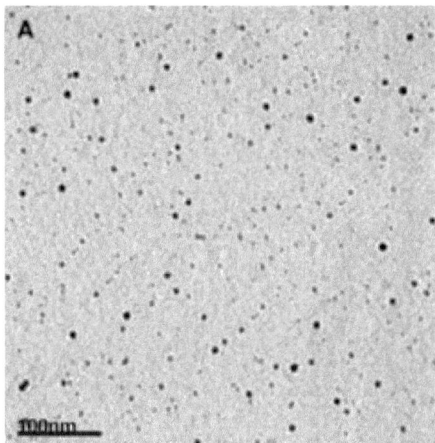

Fig.5. TEM image show Silver nanoparticles
Synthesized by *R.stolonifer*

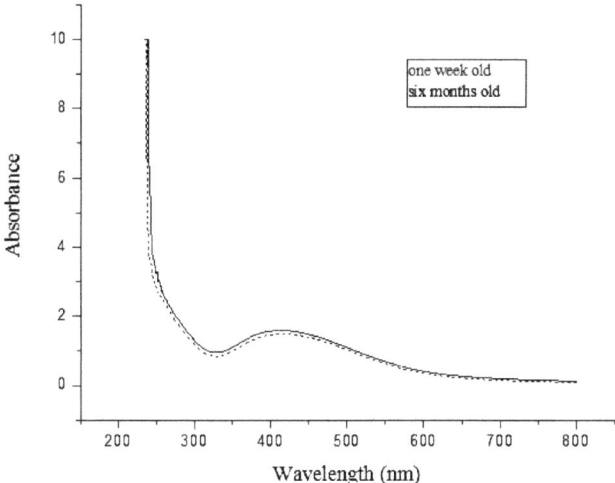

Fig.6. Absorption spectra of silver nanoparticles recorded one week after the synthesis and after six months.

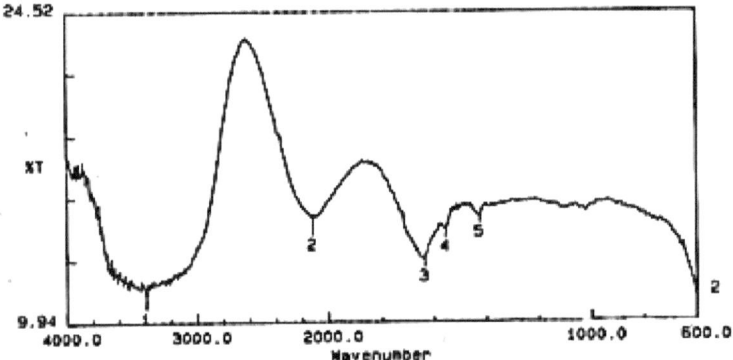

Fig.7.   FT-IR spectra recorded from a drop-coated film of silver nanoparticles
         Synthesized by *R.stolonifer*

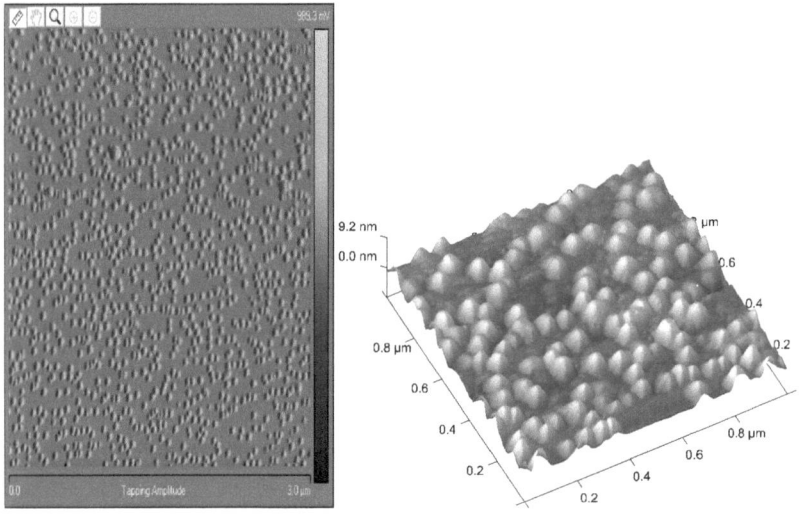

Fig.8a.   AFM picture of the sample   Fig.8b. AFM shows the three dimensional
                                                      image of the silver nanoparticles

Fig.9. Bacteria grown on agar plates at different concentrations of silver nanoparticles (a) *E.coli* (b) *S.aureus*. In each figure the concentration of silver nanoparticles are as follows: upper left, 0 µg/ml, upper middle 0.5µg/ml, upper right 1µg/ml, bottom left 1.5µg/ml, bottom middle 2µg/ml, and bottom right 2.5µg/ml. upper left plate, agar without nanoparticles and without bacterial inoculation.

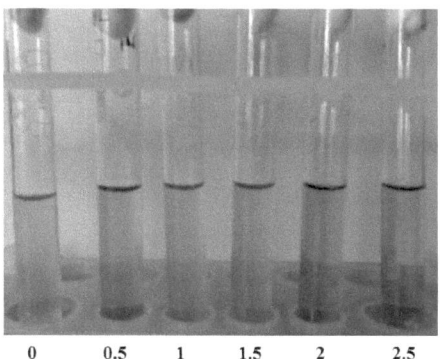

| 0 | 0.5 | 1 | 1.5 | 2 | 2.5 |

Fig.10. MIC of biosynthesized AgNPs against test strains

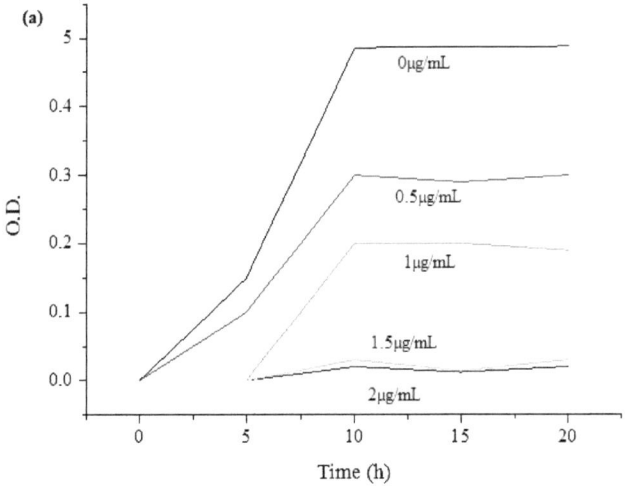

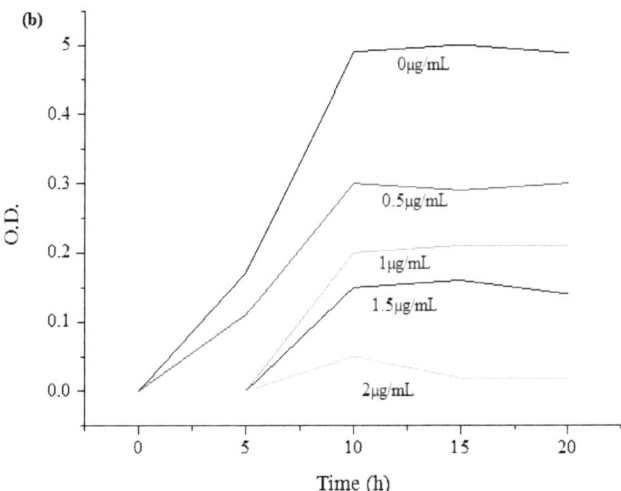

Fig.11. Bacterial growth curve in LB media at different concentration of AgNPs
(a) E.coli (b) *S.aureus*